I0816397

# WHAT IS GLACIOLOGY?

*Glaciology* is the study of snow and ice. It includes the formation, structure, and location of *glaciers*, and how they interact with the **climate**.

The scientists who study glaciers are called **GLACIOLOGISTS.**

# ARE GLACIERS THE KEY TO CLIMATE CHANGE?

DISCOVER THE SCIENCE BEHIND **GLACIOLOGY**
(glay-see-OH-luh-jee)

*Written by Olivia Watson*
*Illustrated by Daniel Limon*

Words that are tricky to understand are in **bold**. Find out what they mean in the glossary.

Words that are difficult to say are in *italics.* Find out how to say them at the back of the book.

In the most extreme, frozen parts of our planet lie glaciers – huge, thick sheets of ice. It may not look like it, but glaciers are always moving! As they move, they scrape against rock, changing shape as chunks of ice break off.

They are so powerful that they can even break and move rocks, creating **valleys** and **mountains,** and completely

**changing the shape of the land!**

Scientists were curious to find out how thick glaciers could be. They decided to drill **deep into the ice!**

The first drills looked different from drills today, and they weren't as strong either. It's difficult to drill glaciers as the ice is **denser** the deeper you get. As technology got better, scientists made an important discovery...

Powerful machines could be made to pull out long tubes of ice, called ice cores, from glaciers. Glaciers are made from layers of new snow that have hardened into ice, so looking at deeper layers is like **looking back in time!**

*Glaciologists* discovered that snowflakes form around things like dust, **pollen, volcanic ash**, and air, trapping information inside glaciers for thousands of years.

Ice is really good at protecting things – freezing temperatures stop **bacteria** from **decomposing** materials. It's so good that some lucky people have even found the bodies of huge animals, like ancient woolly mammoths and woolly rhinos, in layers of **underground ice!**

By studying glaciers and the things trapped inside different types of ice, scientists have worked out lots of amazing things about the animals that used to roam the Earth, the plants that used to grow, and even what the climate was like in the past!

Glaciologists know that places today that are green and full of life were once icy landscapes that looked completely different. They have figured this out by studying rocks and how they were shaped.

We now know that Earth goes through warm periods – like we're experiencing now – and cold periods, called ice ages, when the world was much colder and glaciers spread out to cover larger areas. Some ice ages lasted for thousands of years!

Glaciers aren't the only things in nature that tell us about the climate. Rings inside tree trunks hold lots of clues too. Thicker rings show years when the tree grew bigger and faster thanks to good climate conditions...

the **fossils** of pollen trapped in soil and rock reveal the climates different types of plants were able to survive in...

and the layers of different types of rock tell us all about the long, fascinating history of planet Earth.

Although **climate change** is normal, ice cores have shown that in the last 200 years, the climate has been changing much faster than usual. It started when humans began burning **fossil fuels** to power factory machinery and vehicles like cars. This has made the climate warmer, and is something scientists call **global warming**.

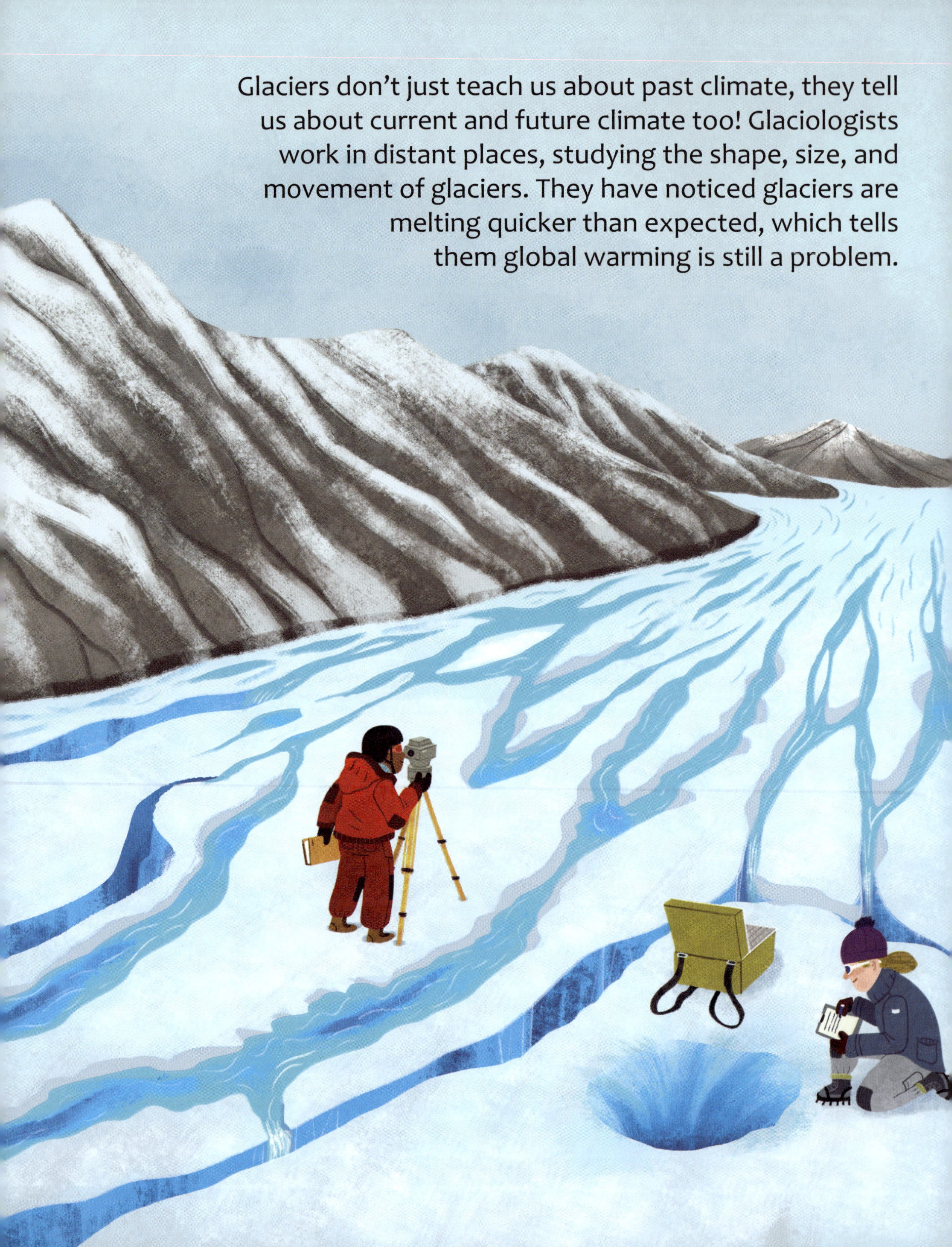

Glaciers don't just teach us about past climate, they tell us about current and future climate too! Glaciologists work in distant places, studying the shape, size, and movement of glaciers. They have noticed glaciers are melting quicker than expected, which tells them global warming is still a problem.

Glaciers usually help keep Earth's climate cool because their white surfaces bounce the Sun's heat back into space. But as global warming is making them melt faster, there is less ice to reflect the Sun's heat. This makes global warming speed up even more! So we need to stop glaciers from melting.

As glaciers melt, the water flows into the oceans, making sea levels rise. This can affect humans because it causes flooding of coastal areas and makes some places **uninhabitable**!

All the extra water can also cause more extreme weather events, like big storms and **hurricanes**.

Melting glaciers make life tricky for animals too. Glaciers cover huge areas of land and are home to many different animals. As glaciers and sea ice shrink, these animals have fewer places to live and hunt, which **puts them in danger!**

Glaciologists are doing really important work. Thanks to these talented scientists, we know that glaciers are the key to understanding climate change, from its past to what we can do about it now and in the future!

Glaciers don't just affect us, they have an impact on **all life on Earth!**

*Slowing down*

# CLIMATE CHANGE

**Climate change may be normal, but it's happening much faster than is safe for humans, animals, and plants. These are just some of the things we can all do to help slow it down.**

## SPEAKING UP

One of the best things we can do is to tell other people about climate change and encourage them to make changes in their lives. The more people that are in the know, the bigger difference we can make!

## GETTING OUTSIDE

Choosing to walk or cycle to nearby places rather than taking a car reduces how much **pollution** is released into the air. Less pollution keeps our air clean and stops it warming up too quickly.

## FLIPPING THE SWITCH

Using less energy is another great way to help. Turning off lights and unplugging electrical devices when they're not being used is easy and makes a big difference!

## THE POWER OF FOOD

It takes lots of resources to grow and transport food. Reducing food waste, and **composting** what can't be eaten, is a big help for the climate.

## PLANTING TREES

Trees are superheroes when it comes to keeping our air clean and the natural world healthy. Looking after forests and planting more trees will help with climate change.

*Ice-cool*

# GLACIER FACTS

**There's so much to discover about the world of glaciology. Do you know the answers to some of the world's biggest questions about glaciers?**

## ARE THERE GLACIERS ON OTHER PLANETS?

Yes! Scientists have found evidence of glacier-like ice forms in many places across outer space, including Mars! Their studies on glaciers in space are only just beginning.

## WHAT MAKES GLACIERS BLUE INSIDE?

The tops of glaciers usually look white, but deeper layers look blue! Deep ice is made of bigger crystals which are packed densely together. They take in red light, but reflect blue light!

## WHAT IS THE THICKEST GLACIER EVER?

Taku Glacier in Alaska is the thickest mountain glacier – it's almost twice the height of the world's tallest building! But parts of the Antarctic Ice Sheet may be even thicker.

## ARE GLACIERS ONLY FOUND AT THE NORTH AND SOUTH POLES?

No! While most are at the poles, glaciers exist on every **continent** except Australia. They've even been found near the **equator**, high up in mountains where it's cold enough for ice to form.

## ARE GLACIERS ONLY STUDIED FROM ABOVE?

No! Sometimes scientists travel inside glaciers to study them. Over time, melted ice under glaciers can wash away huge chunks of ice to form glacier caves!

# GLOSSARY

**Bacteria** – tiny living things that can be found in all natural environments.

**Climate** – long-term temperatures and weather conditions.

**Climate change** – a change in climate over a long time.

**Composting** – turning food waste into usable material.

**Continent** – one of the huge pieces of land on Earth. For example, Africa and North America are separate continents.

**Decomposing** – the process of plants and animals breaking down after they have died.

**Denser** – tightly compacted.

**Equator** – an invisible line that runs around the middle of Earth, an equal distance from the North and South poles. It's usually very hot there.

**Fossils** – the remains or impression of plants and animals that lived long ago.

**Fossil fuels** – fuels that have come from underground, and are made from plant or animal remains.

**Global warming** – the rising temperature of the planet over time.

**Hurricanes** – a type of storm that has strong winds, rain, thunder, and lightning. They form over oceans but can move over land.

**Mountains** – rocky landforms that rise high above their surroundings.

**Pollen** – a fine powder produced by some plants when they reproduce.

**Pollution** – harmful materials that have been released into the environment.

**Uninhabitable** – somewhere that is not safe or suitable for anything to live there.

**Valleys** – long, low areas that are usually found between hills and mountains (see left).

**Volcanic ash** – a powder that comes out of volcanoes during eruptions.

# HOW DO I SAY?

**Glaciers**
GLAY-see-ers

**Glaciology**
glay-see-OH-luh-jee

**Glaciologists**
glay-see-OH-luh-jists

# THE BIG QUESTIONS ANSWERED

**This is more than just a series of books; it is a complete resource. Accompanying each book is a variety of FREE material to engage curious kids with science.**

**www.thebigquestionsanswered.com**

Use the QR code to visit the website, download free resources, and discover other books in the series.

On the website, find out incredible things about glaciologists, including what they do, some of their greatest discoveries, and the people who have made a difference in this field of science.

**The material is also available for home or classroom use, supporting all the information in this book.**

**Teachers' & Parents' Resources**
With discussion prompts and questions, extra information, and facts around key topics.

**Young Glaciologists' Activity Pack**
Fun activities for wannabe glacier experts, including creative writing, drawing, word searches, and much, much more.

**The Big Questions Answered is published by Beetle Books.**
**Beetle Books is an imprint of Hungry Tomato Ltd.**

First published in 2025 by Hungry Tomato Ltd
F15, Old Bakery Studios, Blewetts Wharf, Malpas Road,
Truro, Cornwall, TR1 1QH, UK.

ISBN 9781835691403

A CIP catalog record for this book is available from the British Library.

With thanks to:
Editors: Holly Thornton and Millie Burdett
Senior Designer: Amy Harvey
The team at Beehive Illustration

Information in this book is up to date as of the time of writing.

Printed and bound in China.

Picture Credits:
(t = top, b = bottom, m = middle, l = left, r = right)
NASA: images-assets.nasa.gov/image/PIA13163/PIA13163~orig.jpg 34mr.
Shutterstock: caioacquesta 35mr; chictype Montreal 32bl; DCrane 34bl, 35bl; Sergey Chips 33mr; Rob Pauley 35tl.